OCEAN MAMMALS

BOTTLENOSE DOLPHINS

by Julie Murray

Cody Koala
An Imprint of Pop!
popbooksonline.com

Hello! My name is Cody Koala

This book is filled with videos, puzzles, games, and more! Scan the QR codes* while you read, or visit the website below to make this book pop.

popbooksonline.com/bottlenose

*Scanning QR codes requires a web-enabled smart device with a QR code reader app and a camera.

abdobooks.com
Published by Pop!, a division of ABDO, PO Box 398166, Minneapolis, Minnesota 55439.

Printed in the United States of America, North Mankato, Minnesota.

042025
082025

THIS BOOK CONTAINS RECYCLED MATERIALS

Cover Photo: Getty Images
Interior Photos: Adobe Stock Photos; Shutterstock Images; Getty Images
Editors: Elizabeth Andrews and Grace Hansen
Series Designer: Candice Keimig

Library of Congress Control Number: 2024948410

Publisher's Cataloging-in-Publication Data
Names: Murray, Julie, author.
Title: Bottlenose dolphins / by Julie Murray
Description: Minneapolis, Minnesota : Pop!, 2026 | Series: Ocean mammals | Includes online resources and index
Identifiers: ISBN 9781098247737 (lib. bdg.) | ISBN 9781098248277 (ebook)
Subjects: LCSH: Dolphins--Juvenile literature. | Dolphins--Behavior--Juvenile literature. | Dolphin family (Mammals)--Juvenile literature. | Marine mammals--Juvenile literature. | Aquatic mammals--Juvenile literature. | Sea animals--Juvenile literature.
Classification: DDC 599.533--dc23

Table of Contents

Chapter 1

Intelligent Animals

Bottlenose dolphins are **mammals**. They are some of the smartest animals in the world. They are very good at **reasoning** and problem solving.

The US military uses dolphins to find **sea mines.**

Watch a video here!

Bottlenose dolphins live in warm waters around the world. They are found near coasts but also far offshore. They live in groups called pods. A pod has 2 to 15 members.

Where Bottlenose Dolphins Live

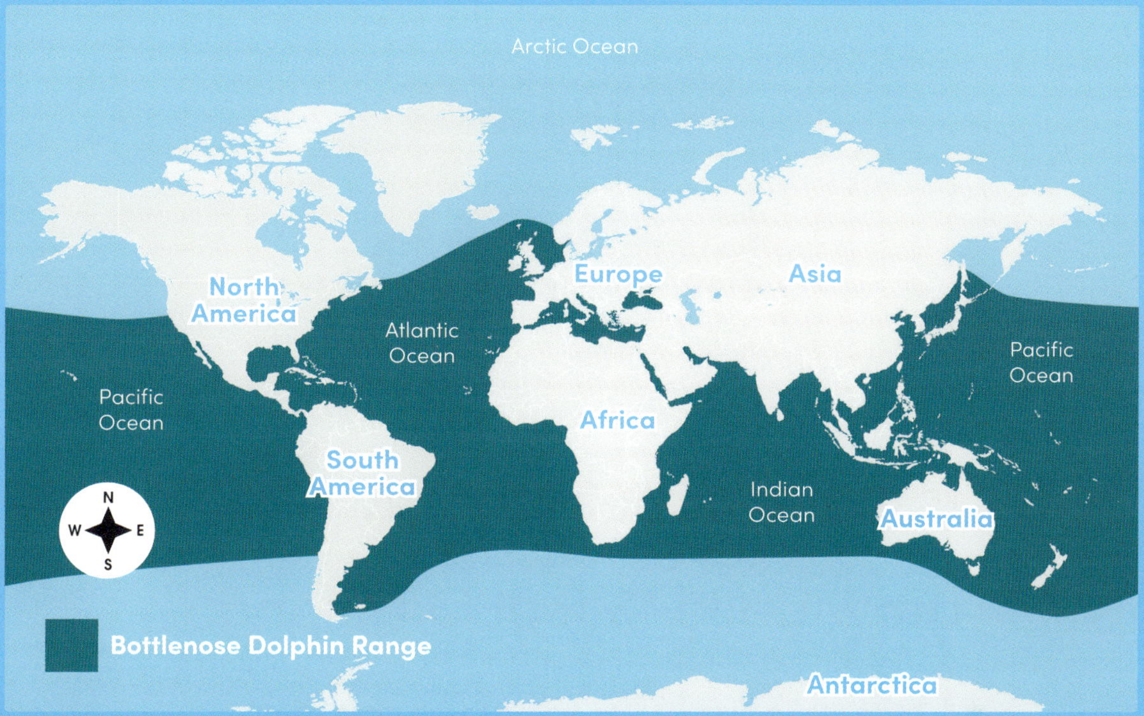

Chapter 2

Smooth Skin

Bottlenose dolphins can grow up to 13 feet (4m) long. They usually weigh around 800 pounds (360kg). Males are larger than females.

Learn more here!

A bottlenose dolphin has smooth skin that is light gray in color. Its back is dark gray, and its belly is white. It has

a short **snout**. A bottlenose dolphin has a large forehead area called a melon.

dorsal fin
blowhole
melon
snout
pectoral fin
flukes

Bottlenose dolphins have tail fins, called flukes, that move them forward. They use their pectoral fins to change direction. The dorsal fin is used for balance. Bottlenose dolphins can swim up to 22 miles per hour (35kph).

Bottlenose dolphins breathe through a blowhole on top of their head. They usually come to the surface two to three

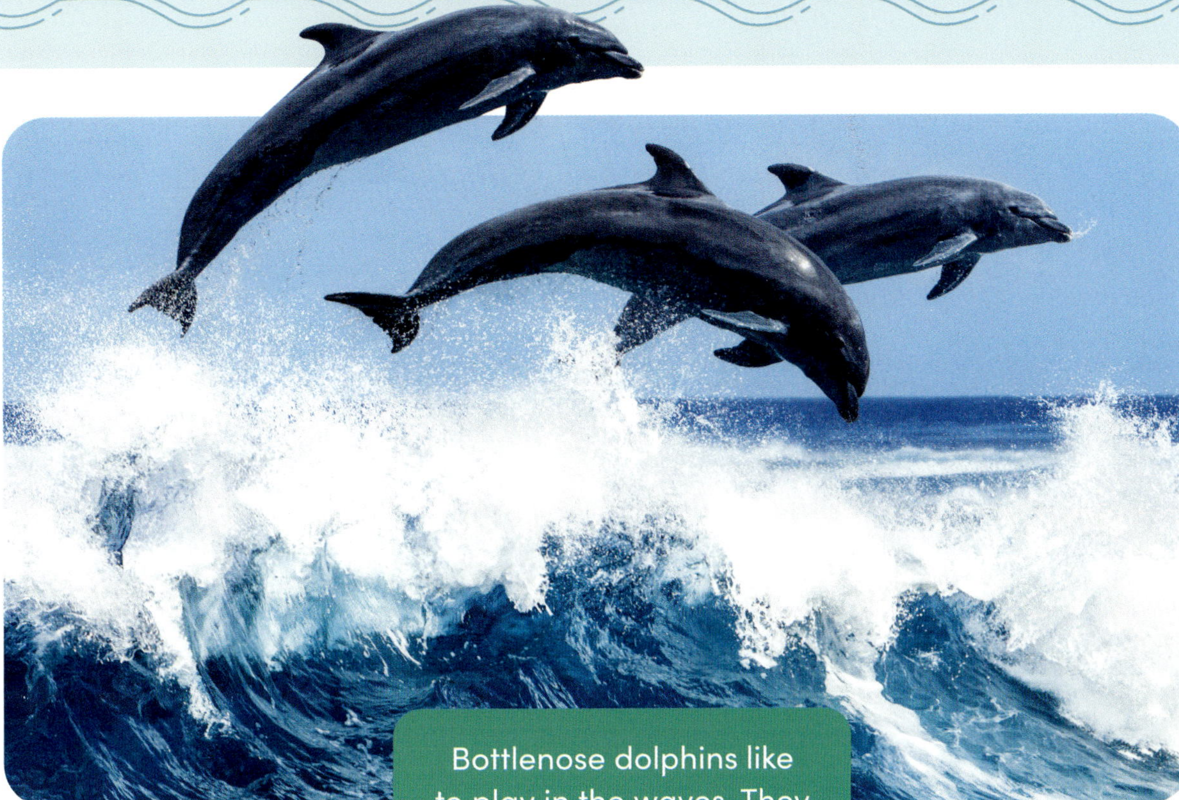

Bottlenose dolphins like to play in the waves. They can jump 15 feet (4.6m) out of the water!

times a minute for air. A skin flap covers the hole when they are underwater.

Bottlenose dolphins use clicking sounds to **navigate** underwater and find food. This behavior is called echolocation. They **whistle** to **communicate** with one another.

Every dolphin has its own whistle that other dolphins recognize.

Chapter 3

Food

Bottlenose dolphins eat fish, shrimp, and squid. They use their teeth to catch food. Then, they swallow it whole. They also use their tail to hit **prey**.

Bottlenose dolphins eat 20 pounds (9kg) of food each day.

Explore links here!

Chapter 4

One Calf

Female bottlenose dolphins give birth to one baby, called a calf, at a time. The calf stays with its mother for about three years. Bottlenose dolphins can live up to 60 years in the wild.

Complete an activity here!

Making Connections

Text-to-Self

Imagine you were a bottlenose dolphin. Would you want to live close to the coast or far out in the open water? Why?

Text-to-Text

Can you think of any books or movies that have dolphins in them? Were they bottlenose dolphins or a different kind of dolphin?

Text-to-World

Bottlenose dolphins are smart animals. Can you name some other animals that are smart? What behaviors make them smart?

Glossary

communicate – to share information with others.

mammal – a warm-blooded animal that produces milk for its young and is usually covered with hair.

navigate – to find a way through.

prey – an animal hunted by other animals for food.

reasoning – thinking through things step by step.

sea mine – an explosive device that is placed in water to damage ships.

snout – the nose, mouth, and jaws of an animal that stick out from its face.

whistle – to make a sharp clear sound by forcing air through a body cavity, such as a blowhole.

Index

Online Resources

popbooksonline.com

Thanks for reading this Cody Koala book!

This book is filled with videos, puzzles, games, and more! Scan the QR codes* while you read, or visit the website below to make this book pop.

popbooksonline.com/bottlenose

*Scanning QR codes requires a web-enabled smart device with a QR code reader app and a camera.